ENERGY WORKS!

Motion

JENNY KARPELENIA

PERFECTION LEARNING®

Editorial Director: Susan C. Thies
Editor: Mary L. Bush
Design Director: Randy Messer
Book Design: Mark Hagenberg, Michelle Glass
Cover Design: Michael A. Aspengren

A special thanks to the following for their scientific review of the book:

Judy Beck, Ph.D; Associate Professor, Science Education; University of South Carolina-Spartanburg

Jeffrey Bush; Field Engineer; Vessco, Inc.

Image Credits:

©Bettmann/CORBIS: p. 6 (right); ©Richard Hutchings/CORBIS: p. 7; ©Jose Lois Pelaez, Inc./CORBIS: p. 10; ©Bettmann/CORBIS: p. 16; ©CORBIS: pp. 20–21; ©Karl Weatherly/CORBIS: p. 22; ©CORBIS: p. 32; ©Paul Almasy/CORBIS: p. 24

ArtToday (arttoday.com): pp. 4, 5, 6 (left), 9, 16 (apples), 18, 31, 34, 35, 36; Corel: cover, pp. 1, 2–3, 8, 10–11 (bottom), 12–13, 14–15, 17, 26 (left), 28–29, 30 (top and bottom right), art on all sidebars; PhotoDisc: pp. 23 (right), 25, 27 (bottom), 26–27 (marbles and balls); Perfection Learning Corporation pp. 10–11 (top), 19, 27 (top), 30 (bottom left)

Printed in the United States of America.

For information, contact
Perfection Learning® Corporation
1000 North Second Avenue, P.O. Box 500
Logan, Iowa 51546-0500.
Phone: 1-800-831-4190
Fax: 1-800-543-2745
perfectionlearning.com

1 2 3 4 5 6 BA 08 07 06 05 04 03

ISBN 0-7891-5866-3

Contents

Introduction to Energy

ENERGY—WHAT IS IT?

Was there ever a time when you felt so tired that you could not even shoot some hoops or play your favorite video game? Maybe you were feeling sick or very hungry. You probably felt as if you had no energy.

Now think of a time when you had lots of energy. You felt as if you could run, talk, ride your bike, or play games forever. Perhaps your parents or teachers even told you that you had *too much* energy.

So what is energy? What forms of energy are there?

Energy is the ability to get things done or to do work of some sort. Anything that accomplishes something is using some form of energy. When you hear the word *work*, do you think of chores around the house? *Work* actually means getting *anything* done. A football sailing through the air is using and giving off energy. A ringing doorbell is using and giving off energy. A shining lightbulb is using and giving off energy.

These examples also show some of the different forms of energy. A thrown football is an example of motion energy. A ringing doorbell is using electrical energy and giving off sound energy. A lightbulb uses electrical energy and gives off light and heat energy. Motion, electricity, sound, light, and heat are all forms of energy. These forms of energy affect our lives every day.

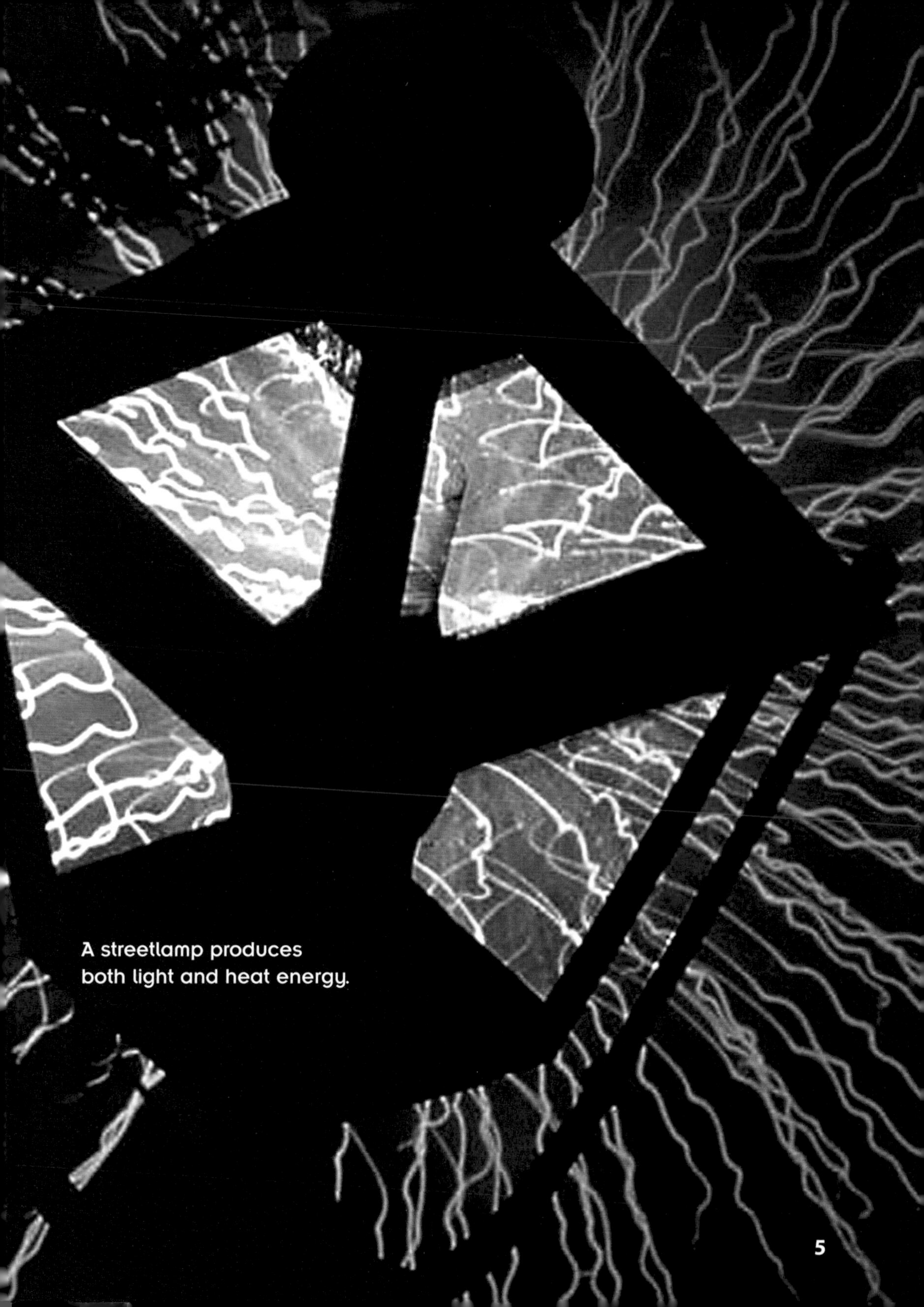

A streetlamp produces
both light and heat energy.

Carnival rides use electrical energy to create motion energy.

Albert Einstein

IT'S A LAW

Scientists perform experiments to test their **theories**, or ideas, about the world. Experiments that produce the same results over and over become scientific laws. One of these laws says that energy cannot be created or destroyed, but it can change from one form to another. This means that all around us, every day, energy is being changed from one form to another. The amount of energy in the universe stays the same, but it is constantly taking different forms.

The famous scientist Albert Einstein was a great thinker. He thought of new ideas that other people had not even imagined. He developed the equation $E = mc^2$. The E stands for "energy." The m stands for "mass" (the amount of "stuff" in an object). His idea shows that energy and mass can change back and forth. So energy can be changed into stuff, and stuff can be changed into energy.

ENERGY WORKS!

Energy is very important. It allows many types of work to be done. People have energy. Plants have energy. The Sun gives off energy. Machines use energy. Read on to find out more about energy, its forms, how it works, and how it is used.

CHAPTER 1

Let's Get Moving!

The lunch crew arrives at school early in the morning to make lunch for that day. The smell of freshly baked rolls fills the hallways. Stomachs growl as it grows closer to lunchtime. When the lunch bell rings, students rush from their classrooms. Each wants to be first in the lunch line. Plastic trays slide. Scoops of mashed potatoes plop. Chocolate milk pours from cartons into messy mouths.

The cafeteria is filled with activity at lunchtime. The constant movement puts everyone in a frenzy. Students move impatiently through the lunch line. The lunch crew quickly moves food onto trays. Forks and spoons purposely move the food into students' mouths. All of these things have energy because of their motion. They all have motion energy.

Motion exists all around you. Clouds float overhead. Birds glide through the air. On the playground, balls bounce and kids run. Cars zoom down the highway. Even when you sit in your seat, you are moving. Your eyes move across the page as you read. Your heart beats. Your blood pumps through your blood vessels. Your breath flows in and out.

The Earth is moving too. It is **revolving** around the Sun at 67,000 miles per hour. The Earth is also **rotating** around its own **axis** at about 1000 miles per hour. It doesn't feel as if the planet is moving, however, because everything on it is moving too.

Life without movement is impossible. This important form of energy guides everything you do, every action you take. Let's move on to the rest of the book to learn more about motion energy!

CHAPTER 2

Potential and Kinetic Energy

POTENTIAL ENERGY

Energy surrounds you, but it is not all being used at the same time. Some energy is being stored for later use. This stored energy is called *potential energy*. *Potential* means "possible" or "having the ability to do something." An object might be resting at the moment, but it is possible for it to move. It has the ability to move if conditions change.

A pencil sitting on your desk has potential energy. It has the potential to roll across the desk or fall off the edge. It is not moving at the moment, but it *could* move. A person could lift or push the pencil. A book could bump into the pencil. A gust of wind could move the pencil. The pencil has the potential for movement.

Some objects gain potential energy because of their location. Any object that is placed off the ground has the potential to fall. The higher an object is, the farther it could fall. Higher objects have more potential energy than lower objects.

Objects may also have potential energy because of the materials from which they are made. When you stretch a rubber band, it has the potential to fly across the room. Rubber is a material that stretches and bounces back easily. It has great potential energy. When you wind up a toy, it has the potential to move. The spring inside stores the energy and then slowly releases it to cause motion. Certain materials are made for movement. Can you think of some others?

KINETIC ENERGY

Energy in motion is called *kinetic energy*. When something causes an object with potential energy to move, the object now has kinetic energy. A pencil falling off your desk has kinetic energy. A paper airplane coasting across the classroom has kinetic energy. If you throw a football at recess, it has kinetic energy. The harder and faster you throw the ball, the more kinetic energy it has.

BACK AND FORTH

Energy is constantly changing from potential to kinetic. Think of a simple game of catch. The ball has potential energy when you hold it in your hand. When you throw the ball, it now has kinetic energy. The ball moves through the air. The ball is then caught by your friend. The ball now has potential energy again as it lies in your friend's glove. This change from potential to kinetic energy happens each time the ball is thrown.

Think of your favorite sports activity. Make a list of all the actions that happen during a game or event. Choose one of the actions, and break it down into potential and kinetic energy. Demonstrate your action for your class, and explain when there is potential energy and when there is kinetic energy.

For example, in the game of volleyball, the server holds the ball while waiting for the referee's whistle. The ball has potential energy. When the server tosses the ball in the air and hits it over the net, the ball has kinetic energy. If the serve is an ace, it hits the ground, eventually stopping. The ball now has potential energy again.

CHAPTER 3

Introducing Force

Everything moves because of force. A force is a push or a pull. When you throw a ball, your arm is pushing the ball. When you walk, your feet are pushing against the ground. When you eat, you are pulling the food up off the table and pushing it toward your mouth. When you fall, the Earth's **gravity** is pulling you down.

Formula 5000 race car

WHAT CAN FORCES DO?

Forces cause objects to change their motion. A force can cause an object to start moving. If you push on a door, it will open. A force can also cause an object to stop moving. If a friend is falling and you catch her, you've stopped her movement by pulling her out of the fall.

A force can cause an object to speed up or slow down. When pushing someone on a swing, you can push harder or softer to control the speed. Or you can pull on the chains to really slow the swing down.

An object can change direction because of a force. If a race car bumps another car on the track, it can cause the car to spin or turn in another direction. The cars are pushing against each other when they collide.

Thunderbird F-16 jets

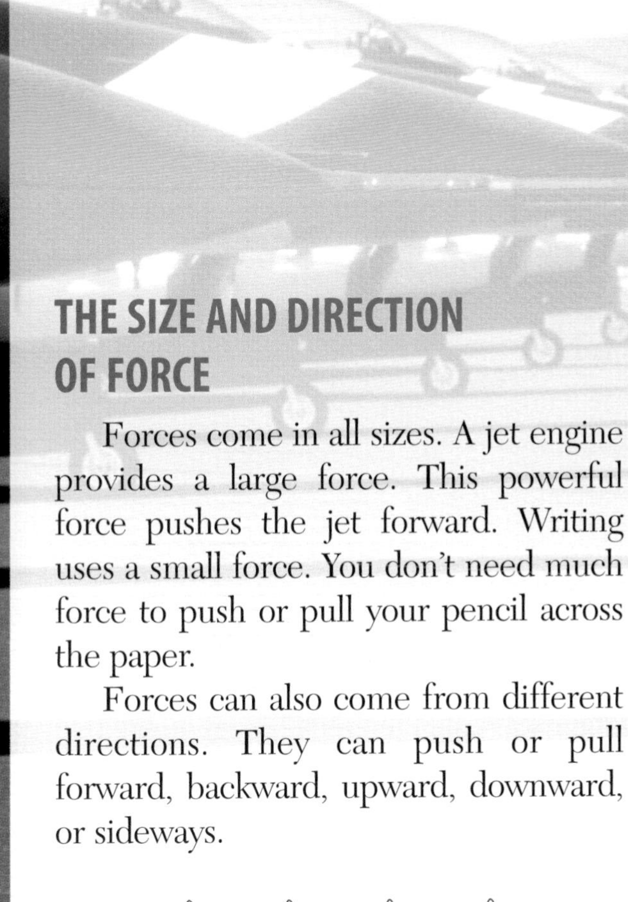

THE SIZE AND DIRECTION OF FORCE

Forces come in all sizes. A jet engine provides a large force. This powerful force pushes the jet forward. Writing uses a small force. You don't need much force to push or pull your pencil across the paper.

Forces can also come from different directions. They can push or pull forward, backward, upward, downward, or sideways.

⚛ ⚛ ⚛ ⚛

Some types of forces have special names. Gravity, centripetal force, centrifugal force, and friction will be discussed in the next few chapters.

CHAPTER 4

Gravity

Gravity is a force that pulls one object toward another object. You have probably already heard the famous story about the falling apple. An apple fell from a tree and hit Isaac Newton on the head. This caused the scientist to investigate gravity. Other people before Newton knew that something pulled things down, but Newton did experiments to explain what gravity is and how it works.

Isaac Newton

Named for Newton

In the metric system, force is measured in newtons—named, of course, for Isaac Newton. In the English system, force is measured in pounds. One pound of force equals about four and a half newtons.

THE DIFFERENCE BETWEEN MASS AND WEIGHT

The amount of material, or "stuff," that makes up an object is its mass. Mass doesn't depend on gravity. You're made up of the same amount of material no matter how hard something is pulling on you. So you have the same amount of mass on Earth as you do on the Moon, even though there's less gravity on the Moon.

An object's weight is a measurement of how hard gravity is pulling on its mass. When you step on a scale, the scale is measuring how much force the Earth is putting on you. The bigger you are (the more mass you have), the more force the Earth uses. So a larger person weighs more. You would weigh less on the Moon than on Earth since the Moon's gravity, or force, is less.

The gravity on the Earth is about six times the gravity on the Moon. You weigh six times more on Earth than you would on the Moon.

Weigh yourself on a scale. Now imagine you're taking a rocket trip to the Moon. How much would you weigh there? Divide your weight on Earth by six to find your weight on the Moon.

HOW DOES MASS AFFECT GRAVITY?

Everything that has mass has gravity. You have gravity. Your pencil has gravity. Even a tiny ant has gravity. These objects don't weigh much, so the force of gravity of these objects is small.

All of the planets in the Solar System have gravity. The Sun and Moon have gravity. These objects have large masses, so the force of gravity of these objects is large.

The more mass an object has, the more gravity it has. The more gravity an object has, the harder it pulls on other objects. All things on Earth have gravity, so they are pulled toward one another. However, the Earth's gravity, or pull, is much greater. So instead of being pulled toward one another, all objects on Earth are pulled toward the center of the Earth.

GRAVITY IN SPACE

Gravity plays a large role in the Solar System. The force of gravity between the Sun, Moon, and planets determines their **orbits**. The Sun has a very powerful force of gravity. Its gravity is so large that it can pull on all of the planets. These planets travel around the Sun. The Sun's gravity keeps them in their orbits.

The Earth is not as heavy as the Sun, so its force of gravity is less. The Moon is even lighter than the Earth. The Earth's gravity pulls at the Moon. It keeps the Moon orbiting around the Earth. The Moon pulls at the Earth too. This causes the ocean **tides**.

As the distance between objects increases, the force of gravity between them gets weaker. Pluto is much farther away from the Sun than the Earth is. So the force of gravity is stronger between the Sun and the Earth than it is between the Sun and Pluto.

Imagine trying to move around on the Moon. Would it be the same as moving on Earth? Would it be easier or harder?

First measure how far you can jump here on Earth. Mark a starting point and take a few practice jumps. Measure each distance. Record your longest one.

Now get back on your imaginary rocket ship. You are headed to the Moon again. This time you want to figure out how far you could jump on the Moon. Multiply your jump distance on Earth by six. This equals your jump distance on the Moon.

You should be able to jump farther on the Moon. The Moon has less gravity, so it pulls on you less. You would jump farther because there would be less force pulling you toward the ground. Imagine what records might be broken if the Olympics were held on the Moon!

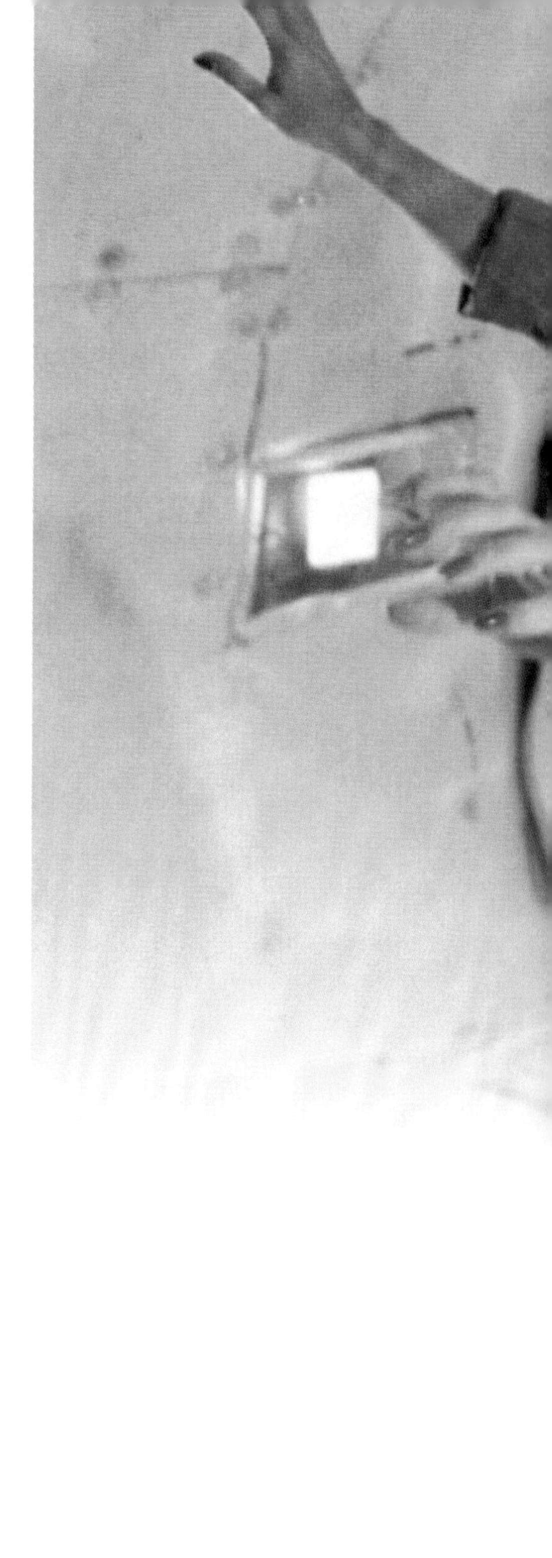

Astronaut Christa McAuliffe participates in weightlessness training for the Challenger Mission.

MICROGRAVITY

When astronauts travel in outer space, they are far from any planet's gravity. They experience a very small amount of gravity called *microgravity*. There isn't enough pull from any planet to keep them **grounded**. This is why people and objects float around when traveling in outer space. Astronauts weigh almost nothing in microgravity. While this might sound like fun, imagine floating in space endlessly. You'd have a difficult time sitting down to watch TV or staying on the ground to run in the school track meet. Perhaps it's time to head back to Earth.

CHAPTER 5

Going in Circles

If there were no forces involved, objects would travel in a straight line. When objects travel in circles, two forces are acting on them.

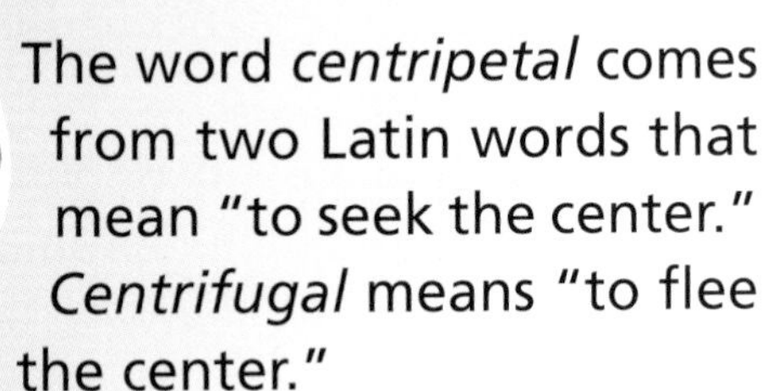

The Meaning Is Central

The word *centripetal* comes from two Latin words that mean "to seek the center." *Centrifugal* means "to flee the center."

CENTRIPETAL FORCE

Centripetal force is the force that makes an object move in a circle. This force pulls an object toward the center of the circle. When the Earth is traveling around the Sun, gravity is the centripetal force keeping the planet on its path. If you tie a ball to a string and spin it around your head, the string is acting as the centripetal force. It pulls the ball inward, keeping it in its path. If you cut the string, the ball would fly off in a straight line.

CENTRIFUGAL FORCE

Whenever an object is pulled to the center of a circle, an equal but opposite force is pushing toward the outside of the circle. This "pushing" to the outside of a circle is known as centrifugal force.

The Sun's gravity is the centripetal force pulling the Earth inward. But at the same time, the opposite is happening. The Earth is pushing back at the Sun with an equal centrifugal force in an outward direction. This keeps the Earth moving around the Sun in a circular path.

Similarly, the string is the centripetal force pulling a twirling ball inward. The ball is pushing back on the string and your hand with an equal centrifugal force in an outward direction. This keeps the ball spinning in a circle.

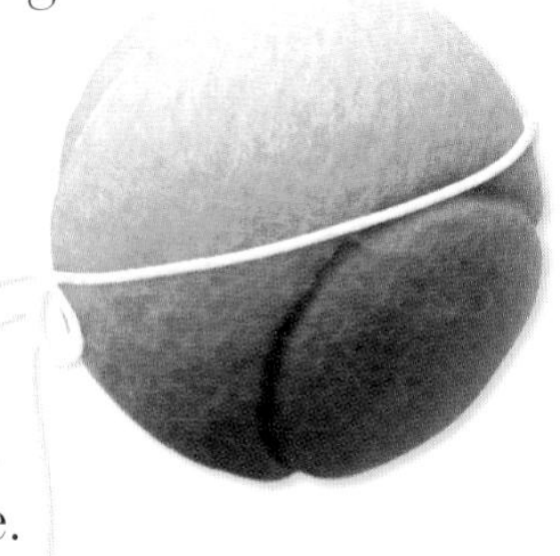

Put a small amount of water into a cup. Now hold the cup in your hand and move it quickly in circles. Watch the water. The water looks as if it is being pushed to the sides of the cup. This leaves the center of the cup without water.

Centripetal force pulls the sides of the cup inward against the water. The water is moving outward due to the equal but opposite centrifugal force.

If you are daring, swing a bucket of water in circles by the handle. Spin the bucket very quickly. Is the centripetal force stronger than gravity? If it is, the water will stay in the bucket. If it isn't, the water will spill out of the bucket as it's pulled toward the ground.

A CENTRIFUGE

Some machines use centripetal and centrifugal force to do their jobs. A centrifuge is a machine that separates solutions into parts by spinning them around. The circular motion will cause the heavier parts to move to the bottom of the container, while the lighter parts remain on top.

Centrifuges have many different uses. Medical labs use centrifuges to separate blood into its different parts. Some businesses separate chemicals using centrifuges. A cream separator is a centrifuge that takes the cream out of milk to make skim milk.

Scientists continue to find ways to use centripetal and centrifugal force to improve our lives.

A worker checks the butter centrifuge on a cooperative dairy farm in Denmark. A cooperative farm is one owned by a group of people who share the work and the profits.

CHAPTER 6

Friction

Motion is caused by forces. Objects can be started into motion by a push or pull. Friction is a force that acts against this motion. It works in the opposite direction of the motion. When you're trying to push something along the ground, friction pushes back on the object. The greater the friction, the harder you have to push to move the object forward.

Friction acts in places where two surfaces touch. There's friction between car tires and the road. When you in-line skate, there's friction between the wheels and the sidewalk. Rubbing your hands together creates friction.

FRIENDLY FRICTION

Friction can be used to create heat energy. When you rub your hands together, the friction creates heat. This is convenient on a chilly day.

Extreme friction can even create fire. Have you ever wondered why rubbing two sticks together can help light a campfire? It's because the friction between the sticks will eventually build up enough heat energy to produce a spark or flame.

Friction also helps you move safely. When you walk or run, your feet touch the ground. Your muscles move your legs forward, while your feet push off against the ground. The friction between your feet and the ground gives you good **traction**. This keeps you from slipping and falling. The same is true when you ride a bike, a skateboard, a scooter, or any other machine that uses friction.

THE SURFACE MATTERS

How much traction you get depends on the surfaces that are rubbing together. Some surfaces are rough and bumpy. Grass, the sidewalk, the road, and carpeting are some examples. More bumps mean more friction. Look at the bottom of your shoes. Do you see grooves and bumps? This rough surface is called the *tread* of your shoes. Your shoe tread increases the friction between you and the ground so you don't slip.

Some surfaces are much smoother. Icy, watery, and waxy surfaces are smooth. Fewer bumps on these surfaces mean less friction. It's easier to fall on these smooth surfaces. There isn't as much friction holding your feet firmly on the ground. If you wear shoes with smooth bottoms, you will also have less friction. So an icy sidewalk and smooth, flat-bottomed shoes are almost guaranteed to end in disaster!

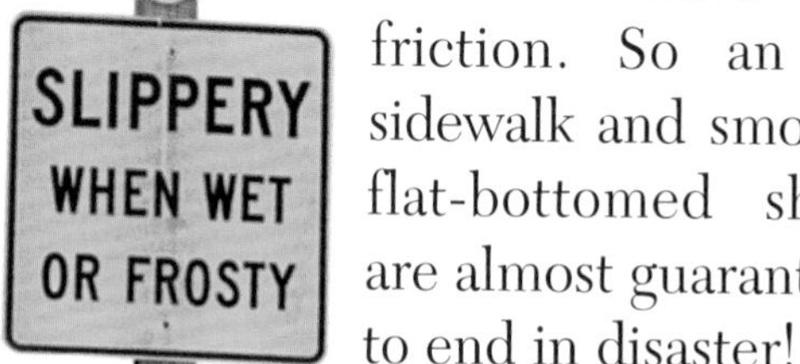

Gather a variety of marbles and small balls. Try to find ones made of as many different materials (glass, metal, rubber, etc.) as possible. Using a ruler or some other slanted surface, roll each of the round objects down your "ramp." Make sure to just release the object at the top of your ramp. Do not give it any additional force. Also, hold the ramp at the same height each time. Measure how far each object travels from the bottom of your ramp. Record your measurements.

Repeat the procedure on at least three different surfaces, such as a tile floor, a hardwood floor, and a carpet. You can also try a sidewalk, grass, or a doormat. Compare your results. Which surfaces seemed to have the least amount of friction? (The objects traveled farthest.) Which had the most? Did the surface of the balls affect the distance they traveled? Write a brief conclusion about your experiment.

FRICTION SLOWS YOU DOWN

Friction is also helpful in slowing down moving machines. Brakes on a bike work because of friction. Squeezing a brake lever on the handlebars starts the braking process. A metal wire connected to the lever is pulled and tightened. This wire is attached to a brake shoe, which looks like a claw. When the brake lever is squeezed, the claw squeezes together. Rubber pads on the ends of the claw create friction. They rub against the metal rim of the wheel. This stops the wheel from turning. The bike slows down and eventually stops.

Car brakes work in a similar way. Stepping on the brake pedal causes a brake pad to rub against the car's metal disks. The friction stops the wheels from turning.

Painful Friction

Friction is also responsible for the scrapes you get when you fall off your bike or take a nosedive off your skateboard. When you fall, your skin scrapes against the ground. The friction between your skin and the ground rubs away your skin. Ouch!

FRICTION MATTERS TO MACHINES

Machines are designed to make work easier for us. Friction can interfere with a machine's **efficiency**. Many types of machines have moving parts. These moving parts may touch other parts or areas of the machine. The friction between the rubbing parts slows down the machine. Eventually the friction causes wear on the parts. For example, the brake pads on a bike or car can wear down until they are too thin to create enough friction to stop the wheels. The pads must be replaced before this happens.

Besides replacing worn parts, oil or grease can be used to reduce the friction between machine parts. The oil or grease makes the parts slide past one another easily. A car's engine needs oil so the **gears** turn smoothly. A door hinge needs oil to keep it from rubbing and squeaking. A bike chain has lots of small moving links. Oil lessens the friction between these links.

When friction is reduced in machines, a smaller amount of energy is needed to make them work. Less friction equals better efficiency.

Engine parts are tested to make sure they are working efficiently.

SHAPE MAKES A DIFFERENCE

Friction doesn't just slow down objects moving along the ground. Friction also pulls against objects moving in water and air.

Water pushes against fish, dolphins, and whales as they swim. This friction slows them down. Their bodies are designed, however, to help them move as fast as possible. These water creatures have a streamlined shape. *Streamlined* means "designed in a smooth, curved shape." The sleek, rounded surface allows the water to move around the animals easily, without creating much friction.

Submarines and boats also have a streamlined shape. This helps them move through the water as efficiently as possible.

Air also pushes against moving objects. Birds, planes, and rockets have streamlined shapes to help reduce this friction.

Cars traveling along the ground also face air friction. Car manufacturers have created streamlined cars to cut down on the effects of this friction. Smaller cars with smooth curves are more streamlined. The air can glide around these cars, increasing their efficiency.

Ferruginous hawk

It's a Bird! It's a Plane! It's Aerodynamic!

Aerodynamics is the study of forces and their affect on motion. If something is aerodynamic, it has been designed to overcome the effects of forces.

CHAPTER 7

How Fast?

Cheetahs are the fastest-moving animals over short distances. They can run up to 70 mph.

"On your mark, get set, GO!" The runners are off. Whoever crosses the finish line first wins the race. The runner with the fastest speed takes home the victory.

SPEED

What is speed? Speed is a measurement of how far an object moves in a certain amount of time. If the winner ran 5 miles in 1 hour, then his speed was 5 miles per hour. If a car travels 55 miles in 1 hour, then its speed is 55 miles per hour.

Desert tortoise

Measure out 30 feet (10 yards). Mark your starting and ending points. Now walk at your normal speed from the starting line to the finish line. Time how long it takes you. Divide the distance you traveled by the time it took you. This is your speed.

Now try it again walking as fast as you can. You can also try running, hopping, or biking. Make a chart of your different speeds.

Everything moves at its own speed. Some things move quickly, like cheetahs, race cars, airplanes, and rockets. Other objects are known for their slow speeds. Snails, tortoises, and maple syrup are known to be slow movers. Did you know that light travels faster than anything? Light has a speed of 186,000 miles per second. Now that's fast!

ACCELERATION

If a runner in a race starts out slowly and then picks up speed as he runs, he is accelerating. Acceleration is a measurement of how fast an object is increasing speed. When you ride your bike, you start out slowly. As you pedal harder and harder, your speed increases. You are accelerating.

When an object in motion decreases its speed, it's called *deceleration*. When a driver is approaching a stoplight, she steps on the brake pedal. This causes the car to slow down, or decelerate.

What if an object is dropped or falls from something without any additional force being applied? In this case, the force of gravity causes the object to accelerate, or speed up, as it falls. The longer an object falls, the more time it has to speed up. An object that falls from a higher place has more time to speed up than one falling from a lower place. So an apple falling from a skyscraper will reach a faster rate of acceleration than an apple falling from a countertop.

Gather a variety of items of different weights, such as a book, a pencil, a baseball, and a shoe. Drop two of the items from the same height. Which one hit the ground first? Try again with another two objects. Repeat until you've tried a few different combinations. Record your observations.

Now try it again using a feather or a sheet of paper as one of the objects. Which item reached the ground first? Why do you think that is?

Read about Galileo and his theory of acceleration to help you understand your results.

Galileo

GALILEO AND ACCELERATION

Galileo was an Italian scientist who lived about 400 years ago. He is given credit for creating the scientific method. This is the process scientists use to solve problems by doing experiments.

Galileo was interested in learning how things move. He designed experiments to help answer the many questions he had. He wanted to study falling motion but had no way of measuring very fast times. Instead, he rolled balls down ramps to find out how far they moved in certain amounts of time. Galileo learned that an object's weight didn't matter. All of the balls accelerated at the same rate. Through his experiments, he determined that objects of the same shape fall at the same rate of acceleration. Even if one object is heavy and one is light, the two will hit the ground at the same time.

What if you dropped a feather and a marble? What if you dropped a flat piece of paper and a crumpled piece? Those objects would not hit the ground at the same time. Those objects have different shapes. The air will push on the feather more than the marble. There will be more air friction pushing against the flat piece of paper than the crumpled piece.

But if you could drop these objects where there is no air friction, would they hit the ground at the same time? Galileo believed that objects had to be the same shape to fall at the same rate of acceleration. However, scientists have repeated Galileo's experiments with modern-day equipment. Using a **vacuum**, scientists have now discovered that all objects fall at the same rate of acceleration. Without air to push against the shapes, the feather and the marble will hit the ground at the same time. The flat paper and the crumpled paper will also reach the ground at the same time. Astronauts have even tested this theory on the Moon.

CHAPTER 8

The Laws of Motion

Many scientists have studied how and why objects move. Scientists all over the world have experimented with motion. Isaac Newton was one of the most famous of these scientists.

LAW OF GRAVITY

Newton was an English scientist who lived about 350 years ago. He is given credit for discovering the law of gravity. This law says that the more mass two objects have, the stronger the force of gravity between them. The law also says that as objects move farther away from each other, the force of gravity between them gets weaker.

LAWS OF MOTION

Newton also developed the three laws of motion. These laws describe the properties, or characteristics, of motion.

First Law

An object will keep on doing what it's doing unless a force acts on it. This means that if an object is sitting still, it will stay that way until a force pushes or pulls it into motion. It also means that if an object is moving, it will keep moving in a straight line until a force pushes or pulls it to speed up, slow down, turn, or stop. This law is also called the law of **inertia**.

Riding in a car is an example of this law. It is very important that you wear your seat belt in a car because you are moving at the same speed as the car. If the car stops suddenly, your body will keep moving. The force of the brakes or the object the car hit will stop the car, but it won't stop you. Without a seat belt, you will continue to move forward until a force, such as the windshield, stops you. Buckling up can protect you against inertia.

Second Law

Force is equal to mass times acceleration. This means that heavier objects need to be pushed or pulled with a bigger force to get them to speed up or slow down. Lighter objects can be pushed or pulled with a smaller force to get them to speed up or slow down.

Imagine pushing an adult on a sled or pulling an adult in a wagon. Adults have more mass, so you would need a big force to move or stop them. If you were pushing or pulling a child, you would need less force since children have less mass than adults.

Third Law

Every action has an equal but opposite reaction. When your feet push against the ground, the ground pushes against your feet. When jet engines on an airplane push out hot gases, the airplane moves forward. When a squid squirts water out of a tube in its body, the squid moves through the water in the opposite direction. Can you think of more examples of this law?

Motion is a very powerful and useful form of energy. It moves our bodies and our machines. It guides the planets in the Solar System. This important form of energy keeps you on the move!

INTERNET CONNECTIONS AND RELATED READING FOR MOTION ENERGY

Energy Quest
(http://www.energyquest.ca.gov/index.html)
This fun site provides information, stories, news, projects, games, and links to other energy sites. Visit the Gallery of Energy Pioneers too.

How Stuff Works
(http://www.howstuffworks.com)
If you have questions about how anything involving energy (and anything else!) works, this Web site is the place to look. It includes sections on energy and electrical power, light, and many other inventions related to energy.

Physics 4 Kids
(http://www.physics4kids.com)
Physics is a branch of science that relates to energy. This site explores motion, thermal (heat), light, and electric energy.

Energy Information Administration
(http://www.eia.doe.gov/kids/index.html)
Review the definition of energy and its forms here. Then check out the Kid's Corner, Fun Facts, and Energy Quiz.

U.S. Department of Energy
(http://www.eren.doe.gov/kids/)
Dr. E's Energy Lab will teach you about solar energy and energy efficiency. A dog named Roofus shows you his energy-efficient home and neighborhood. Many links to other energy sites can be found here.

The Atoms Family (http://www.miamisci.org/af/sln)
This spooky Web site teaches about different forms of energy through simple experiments.

One World.Net's Kid's Channel (http://www.oneworld.net/penguin/energy/energy.html)
Tiki the Penguin discusses the positive and negative sides of different types of energy sources.

Energy by Jack Challoner. An Eyewitness Science book on energy. Dorling Kindersley, 1993. [RL 7.9 IL 3–8] (5868606 HB)

Energy by Alvin and Virginia Silverstein and Laura Silverstein Nunn. Explains a fundamental concept of science, gives some background, and discusses current applications and developments. Millbrook Press, 1998. [RL 5 IL 5–8] (3111906 HB)

Force and Motion by Peter Lafferty. An Eyewitness Science book on force and motion. Dorling Kindersley, 1992. [RL 7.7 IL 3–8] (5868806 HB)

The Magic School Bus Plays Ball: A Book About Forces by Joanna Cole. On a field trip inside a physics book, Ms. Frizzle's class plays baseball in a world without friction and learns all about friction and forces. Scholastic, 1997. [RL 3.2 IL K–4] (5838001 PB 5838002 CC)

Motion by Darlene Lauw and Lim Cheng Puay. Simple text and experiments describe and demonstrate the principles of motion and include a presentation of Newton's three laws. Crabtree Publishing, 2002. [RL 4.1 IL 2–5] (3396601 PB)

- RL = Reading Level
- IL = Interest Level

Perfection Learning's catalog numbers are included for your ordering convenience. PB indicates paperback. CC indicates Cover Craft. HB indicates hardback.

GLOSSARY

axis (AK sis) imaginary line running through the center of the Earth

efficiency (uh FISH uhn see) ability to get good results with as little cost and waste as possible

gear (gear) wheel with teeth (pointed edges)

gravity (GRAV uh tee) force that pulls one object toward another object

grounded (GROWN did) pulled toward a bottom surface

inertia (in ER shuh) idea that an object will remain at rest or in motion unless a force acts on it

orbit (OR bit) path of one object around another

revolving (ree VOLV ing) moving in a curved path around an object

rotating (ROH tay ting) turning around a center or axis (see separate entry for *axis*)

theory (THEAR ee) belief that has been scientifically tested to the point of being accepted as true by most people

tide (teyed) rising and falling of water on the ocean's surface due to the pull of the Moon on the Earth

traction (TRAK shuhn) ability to "stick" to a surface

vacuum (VAK youm) place with no air or matter in it

INDEX